Sylvester Marsh

Section-Cutting

A Practical Guide to the Preparation and Mounting of Sections for the

Microscope

Sylvester Marsh

Section-Cutting
A Practical Guide to the Preparation and Mounting of Sections for the Microscope

ISBN/EAN: 9783744690331

Printed in Europe, USA, Canada, Australia, Japan

Cover: Foto ©berggeist007 / pixelio.de

More available books at **www.hansebooks.com**

SECTION-CUTTING

A PRACTICAL GUIDE

TO THE

PREPARATION AND MOUNTING OF SECTIONS FOR THE MICROSCOPE

SPECIAL PROMINENCE BEING GIVEN TO THE SUBJECT OF ANIMAL SECTIONS

BY

Dr SYLVESTER MARSH

WITH ILLUSTRATIONS

LONDON
J. & A. CHURCHILL, NEW BURLINGTON STREET

1878

PREFACE.

— ◆ —

IF we glance at any of the numerous maga-
zines devoted either wholly or in part to the
subject of microscopy, we shall hardly fail to
be struck with the numerous queries relating
to SECTION-CUTTING, which are to be found in
its pages. A simple explanation of this wide-
spread want is afforded by the fact that the
use of the microscope has at the present day
extended to (and is still rapidly spreading
amongst) vast numbers of students, who, in
many instances, possess neither the leisure nor
the means to refer for information to large and
expensive text-books. Moreover, were they
actually to consult such works, they would
practically fail to obtain the information of
which they are in need, for the coveted in-
struction is to be found in those treatises
only in a scattered and fragmentary form—no

work with which we are acquainted treating of the subject in anything like a detailed manner. To fill this *vacuum* in the literature of microscopy the present manualette has been prepared. Little claim is made to originality, yet the book is by no means a mere compilation, but the outcome of long and extensive personal experience in the cutting and mounting of microscopical sections. Every process described has been put to the test of actual trial, so that its worth may confidently be depended upon. Many of the little points insisted upon in the ensuing pages will doubtless to the practised microscopist appear superfluous or even puerile ; but a vivid recollection of our own early failures and disappointments assures us that it is just these very *minutiæ* of detail which will be found most serviceable in directing and sustaining the faltering footsteps of the tyro.

St. Helens,

    September, 1878.

CONTENTS.

PART I.

	PAGE
Introduction	9
On Cutting Unprepared Vegetable Tissues	10
On Cutting Unprepared Animal Tissues	12
Preparation of Vegetable Tissues	14
Preparation of Animal Tissues	15
Special Methods of Hardening	19
Cutting Hardened Tissues by Hand	19
Microtome	21
Æther Microtome	25
Section Knife	25
Imbedding in Paraffine for Microtome	27
Employment of Microtome	31
Staining Agents	34
Carmine Staining	35
Mounting Media	39
Mounting in Glycerine	41
Uses of Freezing Microtome	47
Employment of Freezing Microtome	48
Logwood Staining	53
Absolute Alcohol	56
Clove Oil	57
Canada Balsam	58
Mounting in Balsam	61
Finishing the Slide	63

PART II.

	PAGE
Special Methods	65
Bone	65
Brain	68
Cartilage	69
Coffee Berry	72
Fat	72
Hair	72
Horn, etc.	73
Intestine	74
Liver	75
Lung	75
Muscle	76
Orange-peel	77
Ovary	77
Porcupine Quill	78
Potato	78
Rush	78
Skin	78
Spinal Cord	81
Sponge	82
Stomach	82
Tongue	83
Vegetable Ivory	83
Wood	83

SECTION-CUTTING.

PART I.

1. *Introduction.*—Many of the most interesting objects with which the microscopist has to deal, cannot be made to reveal their beauty or minute structure until they have been cut into slices or *sections*, of such a degree of thinness as to render them transparent, and thus permit of their examination by transmitted light, with objectives of varying power. Unfortunately, however, very few of the objects of this class are, in their natural condition, in a suitable state to be submitted to this method of procedure. Some are of such a soft and yielding nature that any attempt to cut them is an utter failure, for in place of a perfect section being obtained, nothing remains upon the knife but a mass of diffluent pulp ; others, again, are of such density as to resist the action of any cutting instrument.

It is evident, therefore, that nothing can be done with such refractory materials until, by subjection to appropriate methods of preparation, they shall have been reduced to such a consistence as to render them suitable for cutting. How this is to be accomplished will depend entirely upon the physical and chemical nature of the substance to be operated upon. As the various objects differ so widely from each other in these respects, so must the methods of preparation suitable to each also vary. It is clear, therefore, that no general directions for attaining this end can be given which would be of any practical value. It is possible, however, and very convenient, to arrange the various objects into groups or classes, to the treatment of each of which certain general rules are applicable ; but there will still remain a comparatively numerous series of objects whose individual peculiarities of structure will demand for them correspondingly special methods of preparation. When such objects come to be spoken of, the particular treatment most suitable to each will also be noticed.

2. *On Cutting unprepared Vegetable Tissues.*— There are some few substances, however, which may with more or less success be cut into sections whilst in their natural condition. Such

objects are to be found in the vegetable world
in certain kinds of leaves and allied structures,
whilst in the animal kingdom they are princi-
pally represented by the various internal organs
of man and the lower animals. Special directions are given in text-books for the preparation
of sections of leaves and similar substances.
For instance, it is recommended to lay the
leaf, etc., on a piece of fine cork, and with a
sharp knife to shave off thin slices, cutting
down upon the cork. Another plan is to place
the leaf, etc., between two thin layers of cork,
and cut through the mass. No method, however, is at once so simple and successful as the
process of imbedding in paraffine. To do this,
it is necessary to make a paper mould by twist-
ing a strip of stout writing paper round a
ruler, and turning-in the paper over the end of
the ruler. This mould, the height of which
may vary from an inch to an inch and a
half, should now be about half filled with
melted paraffine mixture (§ 11), the leaf or other
object plunged into it, and held in position by
small forceps till the paraffine has become sufficiently solidified to yield it a support. More
of the paraffine mixture is now poured in
until the specimen is thoroughly imbedded ;
the whole is to be put away in a cold place
for an hour or so, when the mass will be found

sufficiently firm to be cut with ease. Sections may be made with a razor kept constantly wetted with water, or, if the preservation of colour be no object, methylated spirit may be employed for the purpose. As the subsequent treatment of such sections in no wise differs from that required by those cut in the micro-tome, we shall defer its consideration until that method of section has been described (§ 12).

3. *On Cutting unprepared Animal Tissues.*— For the cutting of fresh *animal* tissues several plans may be followed. Thus, if a section of only very limited area be required, it may be obtained by snipping a piece off the tissue with a pair of bent scissors, which, for this purpose, are so made that the blades are *curved on the flat* (Carpenter). If this be carefully performed it will be found that a large portion of the section (particularly at the circumference) so obtained will be sufficiently thin for examina-tion. If a larger section be desired, an attempt may be made to cut it with a very sharp scalpel or razor, the blade of which whilst in use must be kept *flooded* with water or spirit, the latter of which is to be preferred. Recourse may also be had to *Valentin's* knife. This con-sists of two long, narrow blades, running parallel to each other, the distance at which the blades are held apart, and which, of course,

determines the thickness of the section, being
regulated by means of a fine screw passing
through both blades. A milled head attached
to this screw gives a ready means of opening or
closing the blades, so as to bring them to the
desired degree of approximation. The method
of using the knife is very simple. After having
" set " the blades at the desired distance apart
by means of the milled head, the tissue to be
cut is held in the left hand immersed in a basin
of water. The knife is now steadily and with
a rapid motion *drawn* through the tissue, care
being taken that the cut is made in such a
manner that the blades move from heel to point.
By slightly separating the blades and gently
shaking them in the water, the section at once
becomes disengaged.   After use, the blades
must be thoroughly dried, when they may be
smeared with some oil which does not readily
oxidize.   For this purpose, a very suitable oil is
that known as " Rangoon." Though it has been
deemed advisable briefly to describe the preceding methods of cutting unhardened tissues, it
will be found that for the purposes of the ordinary microscopical student sections so obtained
are of very little value. They are always of very
limited dimensions, seldom of uniform thickness, and often so extremely friable as to render
it very difficult and frequently impossible to

submit them with safety to such further treatment as is necessary to fit them for being mounted as permanent objects. This method of section-cutting, however, is not without its uses, for by its means the medical practitioner is provided with a simple and ready method of roughly investigating the structure of morbid tissues, whilst to the general student it furnishes an easy means of making a cursory examination of certain substances, in order that he may determine whether it be worth his while to subject them to some of those various processes of hardening hereafter to be described.

4. *Preparation of Vegetable Tissues.*—Let us now pass to a brief consideration of the methods usually adopted for preparing the various objects for easy section. In the case of *vegetable* tissues, not only do we, as a rule, find their texture of too great density to be readily cut in their natural condition, but they also contain much resinous and starchy matter, of which it is highly desirable to get rid. In order to do this we first cut the substance (say a stem or root) into small pieces, which are to be placed in water for three or four days, by which time all the soluble gummy matters will have disappeared. The pieces are now transferred to a wide-necked bottle, containing methylated spirit, which, in the course of a few days, will

dissolve out all the resin, etc. Many kinds of
woody tissue are by these processes reduced to
a fit condition for immediate cutting; others,
however, are so hard as to render it necessary
to give them another soaking for some hours
in water, to bring them to a sufficient degree of
softness to cut easily. If the wood (as in some
few refractory cases will happen) be still too
hard for section, a short immersion in warm, or,
if necessary, in boiling water, will not fail
effectually to soften it. The treatment of such
members of the vegetable division as require
peculiar methods, will be found described in
future pages.

5. *Preparation of Animal Tissues.*—*Animal*
tissues differ from one another so greatly, both
in consistence and in chemical composition, as
well as in their degree of natural hardness, that
no general rules can be given which would be
applicable to the preparation of the whole class.
Such as are of any considerable degree of hardness, as horn and kindred structures, must be
treated much in the same manner as the denser
varieties of wood, viz., by more or less prolonged immersion in water—cold, hot, or boiling. Those which are of extreme hardness,
as bones and teeth, can be cut only by following certain special methods, full details of
which will be found in the Second Part of

this work (§ 26). Many, and indeed the vast majority of animal tissues, offer a direct con-trast in point of hardness to those we have just been considering. All the internal organs of the body are, when freshly removed, of much too soft a nature to permit, when in their unprepared condition, of easy or perfect cut-ting. It is upon bringing them to that critical degree of hardness, which is often so difficult to attain, that the chief secret of successful section-cutting depends; for unless the hard-ening process has been carried up to, but not beyond, a given point, which varies with differ-ent tissues, the operator, however dexterous, will fail to obtain satisfactory sections. For, if the hardening has fallen short of this critical point, he is, to some extent, in the same position as if he were dealing with unhardened tissues; whilst, if this point has been exceeded, the tissue will have become so brittle as to crumble before the knife. For the purpose of hardening animal tissues, the student has at his command two principal agents, namely, alcohol and chro-mic acid, each of which possesses advantages of its own, but the use of each of which is also attended by its own inconveniences. Thus, by the use of alcohol, there is very much less risk of overhardening the specimen than if chromic acid had been employed. Alcohol, however,

though a capital indurating agent in some
instances, does not answer so well in many
others. Chromic acid is, therefore, to be preferred for general use. It is, however, a very
delicate agent to manage, for unless the greatest
care be taken it is exceedingly likely to overharden tissues submitted to its action, and
when this happens the specimen becomes
utterly useless for cutting, as there is no known
means of removing the extreme brittleness
which it has acquired. By taking the precautions now to be given, this overhardening may
generally be avoided. Let us harden a portion
of some viscus, say the kidney, for instance.
Suppose we cut from the organ five or six small
pieces (from half to three-quarters of an inch
square, *not larger*). These must be placed in a
mixture of equal parts of methylated spirit and
water for three days, at the end of which period
they may be transferred to a solution of chromic acid, made by dissolving twenty grains of
the pure acid in sixteen ounces of distilled
water. The solution should be kept in a wide-
necked bottle furnished with a glass stopper.
At the expiration of seven days, pour off the
solution and replace it by fresh. At the end
of another week, carefully examine the immersed tissues, and by means of a sharp razor
see if they have acquired the necessary degree

B

of hardness to allow of a section of *moderate*
thinness being made. If so, remove the pieces
and put them into a stoppered bottle containing from six to eight ounces of methylated
spirit. If, however, the hardening be found not
to be sufficiently advanced, the chromic acid
solution is to be poured off and again replaced
by fresh. It will now be necessary to examine
the tissues at intervals of about two days, until
they are found to be sufficiently hard, when
they must be transferred to the spirit. Under
no circumstances, however, should they be permitted to remain in the chromic acid longer
than the end of the third week, and though
they should at this time appear not to have
undergone sufficient induration, yet it will be
advisable to transfer them to the methylated
spirit, which in a short time will *safely* complete the process of hardening, without any
risk being run of the tissue becoming ruinously
brittle. It will be noticed, that when the specimens have been transferred to spirit, the latter
will in a day or two become of a deep yellow
colour, whilst a thick, flocculent deposit falls to
the bottom of the bottle. The tissues should
then be removed, the bottle emptied and well
washed, and, being refilled with clean spirit,
the preparations are again to be replaced. This
may occasionally be repeated, until the spirit

becomes and remains perfectly bright and clear. The specimens are then ready for section.

6. *Special Methods of Hardening.*—The brain (§ 27), spinal cord (§ 43), liver (§ 34), and several other organs, etc., require special methods of hardening, details of which will be found in the paragraph devoted to each. In the case of *injected* preparations, the best plan is to harden them in alcohol from the outset, beginning with weak spirit, and gradually increasing the strength as the hardening proceeds. When the object has been injected with Prussian blue, a few drops of hydrochloric acid should be added to the alcohol to fix the colour.

It may here be observed, that specimens of *morbid tissues* require, as a general rule, a shorter immersion in chromic acid solution than healthy tissues do. A very small degree of overhardening speedily renders them brittle and useless. They should, therefore, be removed from the acid medium at the end of ten days or a fortnight, and their further hardening carried on by means of alcohol.

7. *Cutting Hardened Tissues by Hand.*—Our material being now reduced to a fit condition for cutting, let us proceed to consider the several methods by which this may be effected. The readiest and most simple plan, if the piece

be large enough, is to hold it in the left hand, and, having brought the surface to a perfect level by cutting off several rather thick slices, endeavour to cut a thin section by the aid of a very sharp razor, the blade of which must be kept well *flooded* with spirit. As in the use of *Valentin's* knife, so here, great care must be taken steadily to *draw* the blade across the tissue, every effort being made to avoid *pushing* the knife, else the section will be *torn off*, instead of being *cut*. Though this method is of very great importance for many purposes, yet a considerable degree of manipulative skill is required to enable the operator to obtain anything like perfect sections by its means, and, unfortunately, this skill is acquired by very few persons indeed, even after much practice. If the piece which it is desired to cut be too small to be conveniently held in the hand, it may be imbedded in paraffine in the manner already described (§ 3). A very simple imbedding agent, and one of the greatest practical value, is a strong solution of gum arabic, which, upon being dehydrated either by ordinary drying or the action of alcohol, soon acquires such a degree of hardness as to permit it (with the imbedded tissue) to be cut with ease. As this method of imbedding, however, is most frequently resorted to where, by its means, special difficulties have to be overcome, a full description of the process (§ 35) will be deferred until such special cases come to be spoken of.

8. *Microtome.*—Although the preceding plans may be sufficient to answer all his requirements, if the student wishes to obtain only one or two sections of small dimensions, of a given object, if he requires a number of such sections he will find these methods fail him, for even though by practice he may have attained to considerable aptitude in the use of the knife, it will still unquestionably happen that the vast majority of his sections will be more or less imperfect. If, therefore, it be desired to procure a number of perfect sections, of equable thickness and large area, it is absolutely necessary to resort to the use of some form or other of microtome, or section-cutter. This instrument, in its simplest form, merely consists of a stout brass tube closed at one end, and being by the other fixed at right angles into a smooth plate of metal. A plug or disk of brass, accurately fitting the interior of the tube, is acted upon by a fine threaded screw piercing the base of the tube, and by means of which the plug, and any object it may support, can be elevated at pleasure. The object by this means being made gradually to rise out of the tube, sections are cut from it by simply gliding a sharp knife along the smooth cutting plate,

and hence across the specimen. Any intelli-
gent worker in brass would make an instrument
of this kind at a very small cost, and although
perhaps it might lack the finish of an instrument bought at the optician's, it would, if
accurately made, do its work as well as the
most complicated and expensive. If, however,
the student resolves to purchase a microtome,
there are a variety of forms in the market from
which he may choose. A few hints may
perhaps be of service in enabling him to make
a judicious selection. At the outset we may
say that unless the student intends to devote
himself solely to the production of sections of
wood, etc., he ought not to procure one of those
forms of microtome known as wood section-
cutters, in which the object to be cut is held in
position in the tube by means of a binding
screw which pierces its side. Although these
machines are all that can be desired for cutting
hard bodies, they are not so suitable for soft
ones. The chief points to be attended to in
selecting a microtome are, (1) that the cutting-
plate of the instrument be made of glass, or in
default of this, of very hard metal of the most
perfect smoothness; (2) that the diameter of
the tube be neither too large nor too small—it
ought not to be less than ⅝-in., or greater than
1 inch; (3) that the screw, which should be
fine and well cut, be provided with a graduated

head ; (4) that there be some kind of index by which fractional portions of a revolution of the screw may be measured ; and (5) that the plug fit the tube of the microtome so accurately that when melted paraffine, gum, or other imbedding agent be poured into it, it may not find its way between the plug and side of the tube (§ 18). It often happens in cutting tissues imbedded in paraffine, that the pressure of the knife causes the cylinder of the imbedding agent to twist round in the tube of the machine, and so cause considerable difficulty and annoyance. This evil is usually met by running a deep groove across the upper surface of the plug, and into this the paraffine sinks, and so is prevented from rotating. It will be found, moreover, that another difficulty of a kindred, though much more serious character, will frequently be encountered. During section the paraffine has a tendency not only to rotate, but also to become loosened from the subjacent plug, and to *rise* in the tube of the microtome. When this happens the power to cut sections of uniform thickness has completely gone, for some will now be found to be many times thicker than others ; in fact, the irregularity in this respect soon becomes so monstrous as to render it useless to prolong the sitting. In the ordinary run of microtomes no provision seems to have been made to meet this difficulty, and for this reason many instruments, of otherwise great merit, have their efficiency seriously impaired. Fortunately, this imperfection is easily remedied, all that is required being that the upper surface of the plug should be furnished with some kind of projection, having at its summit a table-like expansion, as shown at A in the Figure. The imbedding paraffine, by penetrating beneath and around this, becomes firmly attached to the plug, and thus all risk of its rising is effectually avoided. If the student wishes to secure a really first-class instrument, none can be so

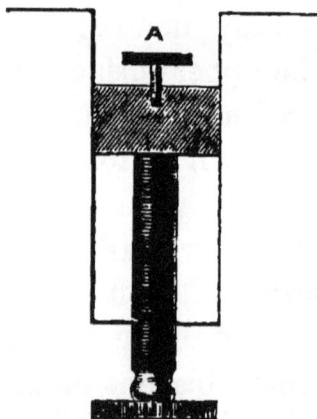

SECTION OF MICROTOME-TUBE SHOWING ARRANGEMENT (A)
TO PREVENT "RISE" OF PARAFFINE.

confidently recommended as the freezing microtome of Professor Rutherford. In addition to its being the best instrument for carrying out the freezing method (§ 18), this machine is equally effective for cutting tissues imbedded in paraffine, or any of the other agents used for that purpose; indeed, whatever work a microtome *can* do, *this one* will perform.

9. *Æther Microtome.*—A word here as to freezing microtomes, where the agent used is æther. Such as have fallen under our notice have not answered the expectations we were justly entitled to form of them. That it is possible to freeze a piece of tissue by their use is undeniable, but it is, as a rule, at an expenditure of such a quantity of æther (only the very best of which must be used) as to constitute it a very expensive proceeding. Another serious disadvantage they possess is, that if the supply of æther be intermitted for only a very short time, the already frozen tissue thaws with great and most inconvenient rapidity.

10. *Section-Knife.*—Of not less importance than the microtome is the section-knife, to be used in conjunction with it. How perfect soever the former, and whatever the dexterity of the operator, unless he be provided with a suitable and well-made knife, he will never succeed in obtaining satisfactory results. As to the most desirable *size* of the knife, much difference of opinion seems to exist, section knives varying in this respect from a blade of extreme shortness to one which fell under our observation, in which the portentous length of

thirteen inches was attained. What advantages were to be expected by prolonging the blade to this extravagant length, must remain an inscrutable mystery to all save its designer. Concerning the *shape* of the knife, it is frequently advised that the surface which has to glide along the cutting-plate of the microtome should be ground *flat.* A most unsuitable arrangement, as a very little actual experience of section-cutting will speedily demonstrate. After many unsuccessful attempts to obtain a really good and reliable section-knife, we determined to have one specially made, which, as it has proved everything that could be desired, merits a brief description. It is of the utmost importance that the blade be made of good and well-tempered steel, not only that it may be capable of receiving an edge of the most exquisite keenness, but also that it may *retain it.* The knife of which we speak (and which was made by Mr. Gardner, of South Bridge, Edinburgh) is furnished with a blade *four inches* long, and ⅞-inch broad, set into a square handle of boxwood, also four inches in length. The thickness of the blade at the back is not quite ¼-inch, whilst *both* of its surfaces are slightly hollow ground. It is essentially necessary that the back and edge of the blade be strictly parallel to each other, otherwise the knife, when in use, will have such a tendency to tilt over as to render its management extremely difficult. It is very easy to discover if this condition be fulfilled, for if on carefully laying the flat of the blade upon a piece of level glass, every portion of both back and edge are found to be in close contact with it, the knife may in this respect be considered perfect. Every student who aspires to be a successful section-cutter should provide himself with a good Turkey oilstone, *and learn to use it.* He should also possess a razor strop, as it will be in constant requisition. It may here be remarked that though *razors*, as a rule, are unsuitable for use with the microtome from want of uniformity in the thickness of their blades, yet, if only a small object is to be cut—for instance, a thin root or stem—very good results may be obtained from their use, especially if one of the old-fashioned make, having a thick back and slightly *concave* surfaces, be employed.

11. *Imbedding in Paraffine for Microtome.*— Having described at some length the various instruments necessary for section-cutting, we will now consider how they are to be used. Let us endeavour to cut some sections—say of a piece of kidney—and in so doing we will adopt the "paraffine" method of imbedding.

Ordinary paraffine, however, when used alone, is rather too hard for our purpose. In order, therefore, to bring it to a suitable consistence, it must be mixed with one-fifth its weight of common unsalted lard, a gentle heat applied, and the two thoroughly stirred together. A quantity of this should be prepared, so that it may always be ready when wanted—it is very conveniently kept in an ointment pot or preserve jar, the top of the latter being well covered, to keep out the dust. When it is intended to use this mixture for the purpose of imbedding, only just about the quantity required should be melted; for in doing this it is advisable to use as low a degree of heat as possible, not only to prevent injury to the tissue to be imbedded, but also that the paraffine when cooling may not undergo such an amount of contraction as to cause it to shrink from the sides of the microtome tube. It is therefore a good plan to effect the melting in a water-bath, a simple kind of which, something after the fashion of a glue-pot, would be made for a few pence by any tinman.

The kidney which we are about to cut has, of course, gone through the process of hardening already described (§ 5), and is now preserved in spirit. A small piece, say half an inch square, is selected, removed with forceps, and placed on a bit of blotting paper, when the surface of the tissue will rapidly become dry (*only the surface* must be allowed to dry). It is the usual plan now to proceed at once to imbed it in the melted paraffine. This is a most undesirable step, and gives rise at a later stage of our proceedings to a great amount of trouble and annoyance, for after sections have been cut from a tissue so imbedded it will be found that portions of paraffine adhere to their edges with such tenacity that in the case of many of them there is no effectual method of removing the paraffine, short of soaking the sections in warm æther; a very objectionable proceeding, for though the æther will undoubtedly remove the paraffine, it will also dissolve out any fatty matters which the section itself may naturally contain. All this annoyance may be prevented by subjecting the tissue to a simple preparatory treatment before it is imbedded in the paraffine. For this purpose prepare a very *weak* solution of gum arabic in water—twenty grains to the ounce. Into this, by means of the forceps, dip for a few moments the already *surface-dried* tissue, taking special care not to squeeze it, or the pressure will cause the spirit from its interior to remoisten the surface, which would prevent the gum from adhering. We shall see the value of this a little later on. Remove the tissue from the solution on to blotting paper, when the superfluous gum will speedily drain off, and in two or three minutes the *surface* will have become quite glazy and dry. Having melted some paraffine mixture in the water-bath, the tissue held in the forceps must be plunged for an instant into the heated liquid and immediately withdrawn, when the crust of paraffine with which it is enveloped will promptly harden. Whilst this is taking place we may make ready the microtome. Having by means of the milled head or handle depressed the plug in the tube so as to leave a free opening about an inch deep at its upper end, we must pour in the melted paraffine, which by this time will have become a little cooler, until the cavity be about half filled. The prepared tissue must now be introduced, care being taken to place it in such a position that the sections may be cut in the desired direction. The tissue must, if necessary, be held in position with forceps or a needle point, till the imbedding material becomes hard enough to give it due support. It is here to be remembered that it will not be advisable to place the tissue in the centre of the tube— it will be much more easily cut if placed rather nearer to that edge of the tube which is situated next the operator in the act of cutting. More paraffine is to be slowly added, until the tissue is completely covered; even after this still more should be added, for it will be found that in cooling the paraffine shrinks so as to leave a cup-shaped depression in its centre, whereby portions of the tissue which were previously covered are again laid bare. The best method of preventing this is to use the paraffine at as low a temperature as possible, and to use plenty of it. The microtome, with its contents, must now be removed to a cool place, when the paraffine will soon become solidified. Whilst this is being accomplished we may make our further preparations. The first thing we require will be a large basin, full of freshly filtered water, and provided with a cover. A small beaker of methylated spirit, with a dipping rod or pipette, will also be necessary. We must now see that the section-knife is in thorough order, to ensure which it will be advisable to give it a few turns on the strop. An ordinary razor will also be of service.

12. *Employment of Microtome.*—The paraffine being sufficiently hard, we will clamp the microtome on to the table, and seat ourselves on a chair of convenient height before it. To our right stand the basin of water, razor, and section-knife;—the beaker of spirit to the left, and a cloth on our knee. A few turns of the microtome screw having brought the

paraffine to the surface, a thick slice is to be cut off, and this repeated until the imbedded tissue comes into view. This preliminary work had best be done with the razor, as it is needless to subject our section-knife to unnecessary wear and tear. By a fractional revolution of the screw the tissue is now slightly elevated, and with the pipette held in the left hand, a large drop of spirit is to be let fall upon its surface. The section-knife, grasped firmly but

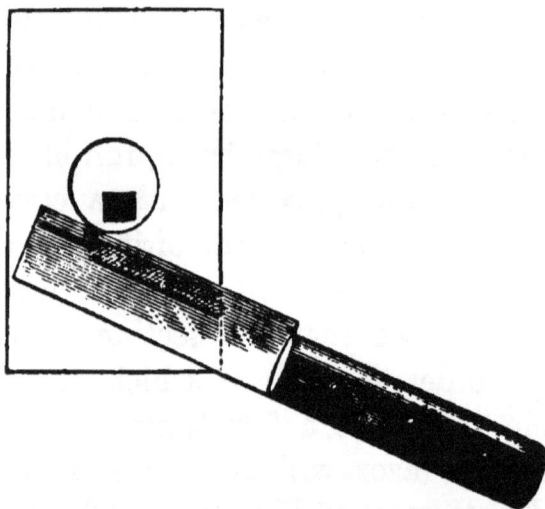

DIAGRAM SHOWING DIAGONAL POSITION OF KNIFE IN COM-
MENCING TO MAKE A SECTION.

lightly in the right hand, is to be laid flat upon the cutting plate of the machine, so as to occupy the diagonal position shown in the figure. Two fingers of the left hand are now laid gently upon the back of the blade, so as to

give it an equable support, whilst the knife
with a rapid motion is pushed in the combined
direction of *forwards* and to the *left*, so that the
blade in cutting the tissue will pass through it
from point to heel. Thus it will be observed
that the stroke of the knife is *from* the operator
—a far easier and more effective mode of cutting than the reverse plan. The blade of the
knife, having the section just cut, either floating in a small pool of spirit on its surface or
adhering thereto, must now be immersed in
the basin of water, when by a little very gentle
agitation of the knife the section will be
floated off. And now we shall find the great
practical value of immersing the tissue in gum
before imbedding, for no sooner is the section
disengaged from the knife than the thin film
of gum which separates the paraffine from it
becomes dissolved, and the section will be
observed gradually to subside to the bottom,
leaving the paraffine floating upon the surface.
After carefully wiping the knife from all shreds
of paraffine, the microtome screw must again be
partially revolved, more spirit applied to the
tissue, and another section being cut, it must
be transferred to the water as before, and so on,
until a sufficient number of sections have been
obtained. As to how thin the sections should
be cut, no general directions can be given ; each

case must be regulated by its own conditions. The denser the tissue, the thinner should the section be; whilst certain substances of loose and spongy texture do not require the sections to be particularly thin—it may be said, however, in a general way that sections, and especially animal ones, *cannot be cut too thin* so long as they remain perfect and entire. If Professor Rutherford's microtome (as made by Gardner) be employed, the head of the screw will be found to be graduated into divisions of slightly unequal value; the sections will therefore be marked by corresponding variations of thickness, so that amongst a number cut, there must be many of the exact thickness to meet the requirements of any individual case.

13. *Staining Agents.*—Before proceeding to mount the sections which have just been cut, it will be very advisable that they should be submitted to the action of some staining fluid, in order to render more clear and distinct their minute structure. Organic substances possess the property of being able to absorb various colouring matters from their solution, and to incorporate such colour into their own texture. This power of attraction is not, however, possessed by all substances indiscriminately, or to an equal extent. Some possess it in a high degree, whilst others appear to be nearly, if not

entirely, devoid of such power. Hence it follows, that if we immerse an organic tissue (one of our sections, for instance) of complex structure, in a suitable staining fluid, the tissue will not become stained in an even and uniform manner throughout, but the several portions of it will receive varying depths of colour in accordance with the varying attractive power of its several constituents. By this means we are enabled in stained sections to discriminate by their difference of shade, minute and delicate structures, which in the unstained condition it would be difficult and often impossible to differentiate. For the purpose of section-staining there are many agents in use, the most generally suitable being carmine, logwood (§ 19), and aniline blue (§ 27); whilst for special purposes chloride of gold (§ 28), pyrogallate of iron (§ 28), and several others are all of much value.

14. *Carmine Staining.* — In the case of animal sections, carmine is, as a rule, to be selected, giving as it does most satisfactory and beautiful results. Tissues may be stained with carmine by two different plans : in the first, a strong solution is used, and the tissue subjected to its action for a very short period only, whilst in the latter only very weak solutions are employed, the time of immersion being considerably prolonged. The rapid method, how-

ever, is not to be recommended, for the strong
carmine acts so powerfully upon the tissue as
to give the various elements comprising it no
time, as it were, to exercise their power of
quantitive selection, but involves the whole
in one uniform degree of shadeless colour. By
adopting the gradual method much better
results are obtained, each portion of the tissue
being now at liberty to acquire its own particular shade. Amongst the various formulæ for
the preparation of carmine fluid, none can be
so safely followed as that devised by Dr. Lionel
Beale. It runs thus :—Place ten grains of the
finest carmine in a test tube, add thirty minims
of strong liquor ammonia, boil, add two ounces
of distilled water, and filter ; then add two
ounces of glycerine, and half an ounce of rectified spirit—this solution ought to be kept in a
well-stoppered bottle. The best vessels in
which to stain sections are small jars of white
porcelain, capable of holding about two fluid
ounces, and furnished with lids—they are much
preferable to beakers or watch glasses, for
owing to the white background which they
afford it is very easy to watch how the staining is proceeding. The carmine solution which
we have just described is both too strong and
of too great density to be used in its pure
state. It will, therefore, require to be diluted
with distilled water before use—the most use-

ful degree of dilution being attained by adding
one part of stain to seven of water. Sections
may be placed in this solution for twenty-four
hours, in which time they will usually be found
to have acquired a sufficient depth of colour.
If, however, the tissue be unusually difficult to
stain, the time of immersion may be doubled,
or still further prolonged, without detriment to
the section.

Having prepared and filtered some of this
dilute solution, say an ounce, let us proceed to
stain with it those sections which we left in
the basin of water (§ 12). Here we are at
once met by a practical difficulty. How are
the sections to be transferred from one vessel
to the other? This is ordinarily effected by
means of a soft camel's-hair pencil. It is a
method, however, open to grave objections, for
the sections so curl around the brush, and get
entangled amidst its hairs, that, notwithstand-
ing every care, valuable sections not unfrequently become torn during transit. Every
difficulty at once vanishes if we substitute for
the brush a small implement, which anyone can
readily make for himself. All that is necessary
is to take a strip of German silver, or copper, of
the thickness of stout cardboard, and about
seven inches in length by five-eighths of an inch
in breadth. The sharp angles are to be filed
off and the edges carefully smoothed, whilst at

a distance of five-eighths of an inch from each
extremity the end must be turned up so as to
form an angle of about 35°. One end must be
left plain, whilst the other, with the aid of a
punch or drill, is to be pierced with five holes
about the thickness of a stocking needle * (see
Figure). If we now dip the perforated end of
this spoon into the water containing the sec-

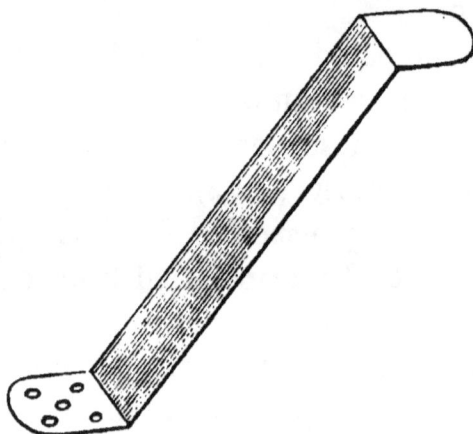

SECTION SPOON.

tions, and gently agitate it, the sections will
rise from the bottom and float about. The

* Dr. Klein describes a kind of "lifter," made by
bending some German-silver wire, but as no drawing
accompanies his description it is not easy to form a clear
idea as to the form of this instrument. In the recent and
philosophical work of Schäfer, a lifter is figured, which
consists of a wire stem, having attached to its end a spade-
like blade. It will be observed that the spoon described in
the text differs from this lifter in having one end perforated, and in this consists the real value of the implement.

spoon is now brought under one of them, and being steadily lifted up the water flows downwards through its apertures, and the section smoothly spreading itself out upon the spoon, may be gently lifted out of the water, and on the spoon being dipped into the staining fluid the section at once floats off. By this simple means sections, however large, thin, or delicate, may with ease be conveyed from one fluid to another, with the utmost certainty of their not being injured during the process. The sections having been in the carmine fluid for about twenty-four hours, as much of the liquor as is possible must be gently poured off, and its place supplied by a freshly-filtered mixture of five drops of glacial-acetic acid to one ounce of water, when in a few moments the carmine will become permanently *fixed* in the tissue, and the process of staining be complete.

15. *Mounting Media.*—The further treatment of the stained sections will entirely depend upon the nature of the medium in which it is intended to mount them. There are a variety of fluids in use for this purpose, the principal being dilute alcohol (§ 26), dammar, or Canada balsam (§ 22-23), and glycerine. These, however, cannot be used indiscriminately, each possessing certain special properties which render it suitable for use with particular

classes of objects only. Thus, weak spirit,
having no tendency to increase the transparency of objects, can advantageously be used
with such only as are already perfectly transparent. It is also more suitable for the preservation of vegetable tissues (when the reten-
tion of colour is no object) than animal, since
with the latter it has a tendency after a while to
cause a kind of granular disintegration, which
ultimately destroys much of the usefulness of
the preparation. Dammar and Canada balsam,
on the other hand, possess very great refractive
power, so that they are of great service in
mounting objects which require their transparency to be much increased. For this reason
they are not well adapted to the preservation
of very delicate or transparent tissues (unless previously stained), the minute details of
which become almost entirely obliterated when
mounted in them. The chief advantage possessed by these resinous media is, that tissues
mounted in them undergo no alteration, even
after the lapse of many years. Glycerine, in
respect of its clarifying powers, occupies an
intermediate position between spirit and balsam, being much more refractive than the
former, infinitely less so than the latter. It
is, therefore, of very great value for the preservation of such tissues as possess a medium

degree of transparency, and which would become obscured if mounted in spirit, or have their outlines rendered indistinct if preserved in balsam. It is of the utmost value for mounting unstained anatomical sections which, when put up in this medium, reveal such minute details of structure as would readily have escaped observation had any other agent been employed. It may also be used with stained sections, but in this case the sections should be of extreme thinness, otherwise the refractive power of the glycerine will be insufficient to render them thoroughly transparent. The great drawback to the use of glycerine is the extreme difficulty experienced in preventing its escape from beneath the covering glass, for it unfortunately possesses such great penetrating power that no cement hitherto devised can be thoroughly depended upon for withstanding its solvent action for any considerable length of time. Attention to the instructions presently to be given (§ 16) will, however, reduce this risk of leakage to a minimum. In the use of glycerine Dr. Carpenter's caution must ever be borne in mind, viz., that, as carbonate of lime is in time dissolved by glycerine, this agent ought never to be employed for the preservation of objects containing such salt.

16. *Mounting in Glycerine.*—To illustrate the method of using this medium we will

mount our present sections in glycerine. In the first place we shall require a deep watch-glass, which is to be half filled with glycerine diluted with an equal amount of distilled water. By means of the spoon, one or more sections may be transferred into this, either directly from the acetic acid solution (§ 14), or if, since cutting, they have been preserved in spirit, they should first undergo a short immersion in a large vessel full of water. The watch-glass should now be covered with an inverted wine-glass, and put away for some hours, in order that the sections may become thoroughly saturated with the dilute glycerine. When this has been accomplished, a slide must be cleaned, and one of the sections, with the aid of the *unpierced* end of the spoon, be transferred to its centre.* As the kind of section with which

---

* The appearance of a slide is vastly improved if the preparation be placed *exactly* in its centre. This may readily be done in the following manner:—Take some very finely-powdered Prussian blue, and rub it up in a mortar with a little of the weak gum solution (§ 11), so as to form a thin blue pigment. A quantity of this should be made, so as always to be at hand. A slide having been cleaned, the *best surface* is to be selected, and on the *reverse* side, by means of the self-centring turn-table, a small circle is to be drawn with a camel's-hair pencil, charged with the pigment. In the centre of this ring, but on the opposite side of the slide, the section is to be placed, when it of course will occupy a position exactly central. When the slide comes to be finished, the blue ring may easily be removed with a wet rag.

we are now dealing is, or ought to be, of extreme thinness, no cell (§ 26) is necessary. After tilting up one end of the slide, so as to drain off as much of the weak glycerine as possible, a drop of Price's best glycerine must, with a glass rod or pipette, be allowed to fall gently upon the section, so as to avoid the formation of air-bubbles. If any of these, however, should be produced, they must be removed with the point of a needle set in a wooden handle,* and the slide then covered with a small bell-glass (or wine-glass). A circular cover is now to be cleaned with a soft handkerchief, and after gently blowing from it any adhering fibres of lint, etc., it will be advisable to hold the side of the glass which is to come into contact with the preparation close to the mouth, and breathe upon it, so as to cover it with moisture. The cover held between the thumb and forefinger of the left hand must now be applied by its edge near to the margin

---

* A *crochet-needle* holder made of bone, and which may be bought at the smallware dealers for about sixpence, makes an admirable handle for microscopical needles. At one extremity there is a small cavity, closed with a cap, for the storage of reserve needles, whilst the other end terminates in a metal tip, provided with a crucial slit and central perforation for the reception of the needle in actual use, and so arranged that, by means of a small screw-nut, needles of various sizes may be firmly held in position.

of the preparation, and the surface of the cover directed in an inclined manner over it. Beneath the overhanging edge of the cover the point of the needle, held in the right hand, is now to be inserted (see Figure). By gently lowering the needle, the cover will come into gradual contact with the slide, driving before it a minute wave of glycerine, in which any air-bubbles that may have become developed are usually carried off.

METHOD OF APPLYING COVER.

A very considerable degree of tact, however, is required to perform this little operation, simple as it may appear, for the retreating wave of glycerine not unfrequently floats out the section, either wholly or partially, from beneath the cover. Air-bubbles, also (the *bêtes noires* of this process), are exceedingly likely to arise. When this happens the best plan to adopt is, by means of the needle point, gently to raise and remove the cover, apply another drop of glycerine to the section, and cover *with a fresh piece of thin glass.* It will now be necessary to remove any superfluous glycerine which may have collected around and near the cover. The great bulk must be wiped away by means of a camel's-hair pencil, slightly wetted between the lips, any remaining stickiness being removed with a bit of blotting paper which has been slightly damped. With a very small camel's-hair pencil, charged with solution of gelatine, a ring must be made round the margin of the cover, of sufficient breadth to take in a small tract of both cover and slide. As this cement is perfectly miscible with glycerine, it readily unites with any of that fluid which may ooze from beneath the cover, and which, in the case of any of the ordinary varnishes, would act as a fatal obstacle to perfect adhesion. To make the cement, take half an ounce of Nelson's opaque gelatine, put in a small beaker, add sufficient cold water to cover it, and allow the mixture to remain until the gelatine has become thoroughly soaked. The water is now poured off, and heat applied until the gelatine becomes fluid, when three drops of creosote should be well stirred in, and the fluid mixture transferred to a small bottle to solidify. Before use, this compound must be rendered liquid by

immersing the bottle containing it in a cup of
warm water. When the ring of gelatine has
become quite set and dry (which will not take
long), every trace of glycerine must be carefully removed from the cover and its neigh-
bourhood, by gently swabbing these parts with
a large camel's-hair pencil dipped in methylated spirit. After drying the slide a ring of
Bell's microscopical cement may be applied
over the gelatine, and, when this is dry, another
coat is to be laid on. If it be desired to give
to the slide a neat and tasteful appearance, it
is a very easy matter, by means of the turn-
table, to lay on a final ring of Brunswick black
or white zinc cement (§ 24). Every care has
now been taken to render our preparation permanent ; but, to make assurance doubly sure,
it will be well to follow Dr. Carpenter's advice,
and, every year or so, to lay on a thin coating
of good gold size.*

* If square covers be employed, they may be fixed to
the slide by a simple method much in vogue in Germany.
A thin wax taper is to be lighted, and being partially
inverted for a few seconds, the wax surrounding the wick
will become melted. After the slide has been freed from
excess of glycerine, a drop of this heated wax is allowed
to fall upon each corner of the cover, and a line of the
melted wax run along the margins of the cover between
these points, so as perfectly to surround it. If a good
coat of white zinc cement be subsequently laid over the
wax a very durable, and not unornamental, line of union
will have been formed.

17. *Uses of Freezing Microtome.*—Our preceding consideration of the method of employing the microtome in conjunction with paraffine
as an imbedding agent (§ 11), will have formed
a very suitable introduction to the study of
the somewhat more complicated process of
imbedding the tissue in gum, for section in
the freezing microtome. This method is of
the utmost value to the practical histologist,
for by its means he is enabled with ease to
possess himself of perfect sections of several
structures, the cutting of which, before the
introduction of this process, was always a
matter of difficulty and anxiety. The freezing
microtome is especially valuable for the section of such substances as from their extreme
delicacy are liable to be injured by being
imbedded in paraffine—for instance, the delicate
villi of the intestines become very frequently,
by the use of paraffine, denuded of their epithe-
lium, and the villi themselves not seldom
become torn off or otherwise damaged. The
great value of the method is also very well
seen in the treatment of those tissues which,
like the lung, are of such loose and spongy
texture as to offer insufficient resistance to the
knife unless their interstices have previously
been filled up with some solid yet easily cut
material. As the space at our command is

strictly limited, we are precluded from entering as fully into this branch of section-cutting. as the importance of the subject demands and our own inclination would lead us. To those who wish to become thoroughly conversant with the full value of this method we cannot do better than recommend the perusal of Professor Rutherford's *Practical Histology*, 2nd edition, than which, on the whole subject of physiological microscopy, no treatise with which we are acquainted is at once so plain, practical, and profound.

18. *Employment of Freezing Microtome.*—A very suitable object with which to demonstrate the method of using this form of microtome will be afforded us by a portion of intestine, say of the ileum of a cat or dog. Suppose we have some of this in methylated spirit—let us select a piece of about half an inch in length. Our first care will be to deprive this of its spirit; for so long as the tissue remains impregnated with alcohol it would, of course, be impossible to freeze it. We will, therefore, throw it into a large basinful of water, and leave it there for twenty-four hours, during which time it would be as well to change the water once or twice. We shall now require a strong solution of gum. This, which should have been made some time previously, may be

prepared by placing a quantity, say three or four ounces, of ordinary gum arabic in a glass beaker, and adding sufficient water to cover it —the mixture must be stirred occasionally with a glass rod until solution has taken place, which will be in a few days. If *necessary* a little more water may be added, but so long as the gum will pour from vessel to vessel it cannot well be made too strong. Mucilage, by keeping, is very apt to become sour and mouldy—this may be prevented by adding to each ounce of the water with which it is prepared about half a grain of salicylic acid. We now pour some of this mucilage into a small vessel—an egg cup will answer very well—and into it transfer the piece of ileum from the water. Here we must allow it to remain for a time sufficient to permit of its becoming thoroughly saturated with the gum, for which purpose some hours will be necessary. When this soaking has been accomplished we will prepare the microtome, which we will assume to be Rutherford's. In the first place it will be necessary to remove the plug—which is to be done by turning the handle connected with the screw until the plug rises so high in its tube that it may be grasped with the fingers and removed, when it is to be well smeared all over with sperm

D

oil and replaced. This is done to prevent any
unpleasant adhesions taking place whilst the
freezing is going on. We must next depress
the plug, so as to convert the upper part of
the tube into a kind of "well" of sufficient
depth to hold our specimen. It will now be
very advisable to look carefully into this *well*,
and observe whether the plug fits accurately
into the tube (§ 8) for if there be any interval
between the two it will give rise to much
subsequent annoyance, as the gum penetrating
this interstice will there become firmly frozen
into irregular patches, which will so interfere
with the even gliding of the plug within its
tube as to cause the former to ascend in such
an irregular and jerky manner as to be utterly
destructive of all accuracy in the cutting. If this
defect be observed, it may be at once remedied
by dropping a small quantity of gently heated
paraffine into the *well*, which will effectually
close up any fissures. The microtome, by
means of its clamping arrangement, must now
be firmly attached to the table, and a suitable
vessel be placed on the floor beneath it, so that
it may catch the water which will issue from
the waste-pipe of the apparatus. The next
requirement is a supply of block ice and finely
powdered salt. A lump of the ice must be
wrapped in a towel, and crushed into small

pieces; these, by means of a large mortar, are to be further reduced to a very *fine powder.* Any attempt to hurry over this troublesome part of the operation will lead to future disappointment, for unless the ice be used in a very fine powder great delay (at least) in the freezing will be the result. With the aid of a small spoon the ice and salt are in alternate spoonsful to be conveyed into the freezing-box of the machine, great care being taken that the cavity under the cutting-plate and around the tube be thoroughly packed, after which the uncovered portion of the box should also be well filled. The *well* is now to be filled with the strong gum to within a little distance of its top, and a piece of sheet gutta-percha (such as shoe soles are made of) being applied over the well, and kept in position by a weight, we must wait until the freezing commences. In a short time we shall notice that the gum has acquired a thick muddy appearance. The tissue must now, by means of the forceps, be transferred to the well, and there placed in such a position that the sections, when cut, shall run in the desired direction. After more gum has, if necessary, been added, so as completely to cover the tissue, the well is again to be covered, and attention given to the freezing-box. As the mixture which this contains

becomes melted, it must constantly be renewed,
care being at the same time taken that the
mouth of the discharge-pipe be kept quite free,
otherwise water accumulating in the box, the
freezing mixture will degenerate into a useless
puddle. When the gum becomes sufficiently
hard to cut, this must be done much in the
same manner as if paraffine had been used
(§ 12). In this case, however, no fluid will be
required, or must be used, to wet the knife
with, and especial care must be taken that in
disengaging the sections from the knife into
the water they be not torn. These sections
often adhere very tenaciously to the blade, but
if a little patience be exercised the water will
soon float them off in safety—much more safely
than if any attempt be made to liberate them
prematurely. There is one circumstance connected with the use of the freezing microtome
which is rather annoying. The moisture of the
breath and atmosphere is apt to become con-
densed on the cutting-plate, and here, mixed
with accidental smears of gum, it becomes
frozen into a jagged and irregular sheet of ice,
which not only seriously interferes with the
smooth play of the knife, but also constitutes a
real peril to its edge. As this evil cannot be
avoided, all we can do is, by constant wiping, to
keep the cutting-plate clean and free from this

accumulation. This is best done with a bit of soft rag *just moistened* with spirit, but this must not come into contact with any portion of the cylinder of frozen gum, else it will instantly thaw it. When using the freezing microtome it is always advisable to wear an apron, otherwise our clothes may receive considerable damage from the constant splashing of the salt water, as it falls from the waste-pipe into the vessel beneath it. After use, the microtome must be well washed in plenty of cold water till every trace of salt be removed, for if any of this remain it will quickly corrode the brass-work of the instrument. The plug and screw, as also the section-knife, should be well smeared with Rangoon oil before the machine is put away.

19. *Logwood Staining.*—The employment of logwood as a staining agent is now becoming very general. It acts much in the same manner as carmine, but the violet colour which it produces is by many thought to be of a more soft and agreeable character than that due to the action of carmine. A valuable and very convenient property also which it possesses is that it stains tissues very rapidly, and this without interfering with that differential kind of coloration (§ 14) upon which the chief value of all staining pro-

cesses depends. A simple method of preparing the logwood fluid is to mix an aqueous solution of extract of logwood with a solution of alum (1 to 8) till the deep impure red colour has become violet, and then to filter the mixture (Frey). This will stain sections in about half an hour. This stain, though here mentioned for the ease with which it may be made is, as a rule, very inferior to a fluid prepared directly from hæmatoxylon, the alkaloid or active principle of logwood. As, however, it is difficult and troublesome to make the solution in this manner,* it will be advisable for the student to purchase, ready prepared, such small quantity of the dye as he may require. Small bottles may be obtained for a few pence of Mr. Martindale, 10, New Cavendish Street, London, and from repeated trials of this solution we can recommend it as producing excellent results. It is a very strong fluid, and requires to be diluted before use. The degree to which the dilution must be carried cannot, however, be very accurately indicated, for all staining fluids of this nature possess the very undesirable property of becoming decomposed by age. After the fluid has been kept for some time, a portion of the

---

* Should the student, however, determine to prepare this solution for himself, he will find a good formula for the purpose in Schäfer's " Practical Histology," p. 176.

colouring matter is thrown out of solution, and
becomes deposited upon the sides and bottom
of the vessel in which it is contained, hence the
older the preparation, the weaker it will have
become. As the time required for staining
with logwood is but short, it is desirable that
all the sections should begin to be submitted
to its action at the same time, otherwise some
will become more deeply stained than others.
A good plan is to fill a small porcelain jar
(§ 14) with filtered water, and into this transfer
the sections. Whilst they are settling well
down to the bottom, a mixture must be prepared of half a drachm of Martindale's solution
(fresh) to one ounce of distilled water, and
everything got in readiness for its immediate
filtration. The water is now very gently to be
poured off the sections, and if care be exercised
this may be done in such a manner as to leave
them undisturbed at the bottom, after removing
almost every drop of water. The diluted logwood fluid must now be *immediately* filtered
upon the sections, so that they may run no risk
of becoming dry. In the present instance the
staining may be allowed to proceed for about
thirty minutes, and this will be found a convenient time for the immersion of the general
run of animal sections. If the logwood fluid
be not quite fresh, either a little more of it
will have to be added to the water, or the time

of immersion must be prolonged until the desired depth of colour has been produced. It is well whilst the staining is going on gently to shake the vessel occasionally, so that the sections may not remain in a heap at the bottom, but all be as fully as possible exposed to the action of the dye. When the staining is judged to be complete, the logwood solution must be gently poured off, leaving the stained sections at the bottom of the jar, when they should be quickly covered with methylated spirit, which will *fix* the colour. We shall now be able to see if the coloration obtained be perfectly satisfactory. If not deep enough, it is very easy again to submit them to the action of the dye for a few minutes longer. If on the other hand, and as more frequently happens, the coloration should be too deep, the excess of colour may readily be removed by transferring the sections for a short time into some diluted acetic acid prepared by adding five drops of the glacial acid to an ounce of water: The action of this should be carefully watched, and when the colour has been reduced to the desired tint the sections may be retransferred to the methylated spirit.

20. *Absolute Alcohol.*—As we purpose mounting the sections which have just been stained, in Canada balsam, we will briefly consider the

preliminary treatment to which they must be submitted before this can be effected. The object of this is to abstract from the tissue all its water, for if any moisture be permitted to remain in the section it will, when mounted in balsam, become obscured and surrounded by a kind of opalescent halo, due to the imperfect penetration of the balsam into the only partially dehydrated tissue. The old-fashioned plan of dehydration was simple exposure to the air. The method now generally adopted is to bring about the same result by means of absolute alcohol. This fluid has such a strong affinity for water that tissues submitted to its influence are rapidly and effectually deprived of any water they may contain. Absolute alcohol in small quantity may be obtained from the druggist at about sixpence per ounce. It will be necessary for the student to provide himself with a little of this agent, say about two ounces, the method of using which will very shortly be explained. Absolute alcohol must be kept in a bottle with a very accurately-fitting stopper, in order to prevent its absorbing moisture from the air. For our purpose such a bottle, having a neck *as wide as possible*, is to be selected.

21. *Clove Oil.*—After being thoroughly dehydrated the sections may, in special instances

(§ 48), be at once mounted in balsam; but, as a general rule, it will be found necessary (particularly in the case of animal sections) to treat them with some clarifying agent, in order to remove the cloudiness and opacity which is (in part) due to their previous immersion in alcohol. For this purpose turpentine, or any of the essential oils, may be used : of these, oil of cloves is to be specially recommended. It is rather expensive, ranging from sixpence to one shilling per ounce; but, as a drop or two will be sufficient for preparing each slide, only a small quantity—say half an ounce or an ounce—need be procured. The most convenient vessel in which to keep the oil is one of the small test bottles used by watchmakers. These bottles are provided with a glass cap to exclude dust, and the stopper is prolonged into a glass rod, which dips into the bottle. The use of this rod and the method of employing the oil will be explained shortly.

22. *Canada Balsam*, as ordinarily met with, is a thick resinous balm of great viscidity, but readily rendered perfectly fluid by the application of heat. Formerly, sections were mounted in this medium in its pure state, but owing to the annoyance which was so constantly being experienced from the tenacity with which intruding air-bubbles were held by the viscous

medium, this plan of mounting is rapidly falling out of use.* It is now usual to employ the balsam in a diluted condition, the two chief diluents being chloroform and benzole. As balsam, however, often contains more or less moisture, it is desirable to drive this off before adding the diluent. A very convenient way of doing so is to expose some pure balsam to the heat of a cool oven for several hours, when the

* Although we cannot too strongly insist upon the use of chloroform-balsam wherever practicable, yet it some-times happens in the mounting of substances of *considerable thickness*, that after all the chloroform has evaporated an insufficient amount of balsam is left behind to fill up the cavity between slide and cover. In such cases, there-fore, it is advisable to use pure balsam, which may be done in the following manner. The object having been previously thoroughly dehydrated by immersion in absolute alcohol, is to be thence transferred to a little good tur-pentine or benzole, where it should remain until perfectly transparent. It is now to be placed in the centre of a slide which has been gently warmed, and a drop or two of *fresh* fluid balsam added, the greatest care being taken to prevent the formation of air-bubbles. Should such arise they must be touched with the point of a heated needle, which will cause them to burst and disappear. The chief difficulty of the process has yet to be encountered in the application of the cover ; for it is during this procedure that the development of air-bubbles is most likely to take place. This annoyance may, however, be entirely avoided by taking the simple precaution of dipping the cover into turpentine before it is applied (§ 16), when it will be found that "you can't get air-bubbles, even if you try." The courtesy of Mr. J. A. Kay, of Chatham, enables us to give our readers the benefit of this practical "*wrinkle*."

balsam will be found to have assumed a hard, vitreous character. It should now be broken into small pieces, these put into a bottle, and some methylated chloroform added, which in a little while will completely dissolve the hardened balsam. More chloroform is then to be added, until a solution is obtained sufficiently thin to run through filtering-paper. A glass spirit lamp must now be procured, having

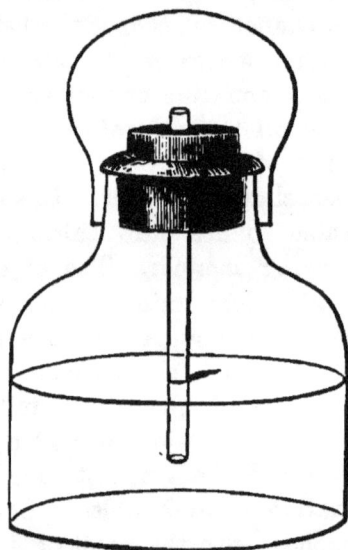

SPIRIT LAMP ADAPTED TO CONTAIN BALSAM.

a capacity of about two ounces, and provided with a cap. Into the wick-holder of this (which must be made of porcelain) a hollow glass tube is to be so fitted that its end dips into the lamp to within about a third of the bottom. (See Fig.) The thin chloroform balsam is now to be filtered into this lamp, very fine filtering-paper—through which a little chloroform has first been passed—being used for the purpose. When the lamp is full it must (deprived of its cap) be put in a warm place until sufficient of the chloroform has evaporated to leave behind it a fluid of the consistence of thin syrup.

23. *Mounting in Balsam.*—Let us now return to our sections which, it will be remembered, were left in methylated spirit (§ 19). These we will now mount in balsam, and although, of course, any number may be proceeded with at the same time, yet, to avoid confusion, in the following directions one section only will be spoken of. This section then is, with the perforated spoon, to be transferred to the bottle of absolute alcohol, where it may remain for about an hour—considerably less time is *actually* required, but as from constant use the spirit becomes weakened, it is as well to be on the safe side. It must now be removed to the centre of a clean glass slip, and here the *plain* end of the spoon comes into use. If this be employed for effecting the transfer, it will be found that when the section is being removed from the alcohol it will bring along with it a small pool of the spirit. A slight touch of the needle applied to the edge of the section will

cause it to float from the spoon on to the slide, at the same time carrying the pool of alcohol with it, in which it will gently spread itself out upon the slide without the faintest risk of injury. The superfluous spirit is now to be drained off, and just as the section is becoming glazed and sodden-looking (*not dry*) we must, by means of the long glass stopper (§ 21), apply to it a large drop of clove oil. The oil, however, should not be placed *on* the section, but be allowed to drop on to the slide near to its margin. By gently tilting the slide the oil will gradually insinuate itself *beneath* the section and slowly ascend through it to the surface. The slide should now be covered with a bell-glass (or wine-glass), and about two minutes allowed for the oil thoroughly to saturate the section. As much as possible of the superfluous oil must then be drained off, and the remainder removed with blotting paper. By means of the glass rod a small quantity of chloroform balsam is now taken from the spirit lamp which contains it, and allowed gently to fall upon the section, which must then be covered with a thin glass circle in the manner previously described (§ 16). When the object is very fragile, it is a good plan, after draining off the clove oil, to apply the cover directly upon the section, and then to place a drop of the balsam near to the edge of the cover. This, by capillary attraction, will speedily diffuse itself beneath the cover, flowing over and surrounding the object, without in the slightest degree disturbing its position. If, during the process of mounting, any air-bubbles arise, we may view their development with equanimity, being well assured that as the chloroform evaporates they too will quickly disappear. When the mounting is completed, the slide should be roughly labelled and placed on a warm mantel-piece for a few days to dry.

24. *Finishing the Slide.*—In the course of two or three days it will be advisable to take an old penknife, and after heating the blade in the flame of a spirit lamp, gently to run the point of it round the margin of the cover, so as to remove any excess of balsam which may have oozed from beneath it. In a few days more, any remaining balsam may be carefully scraped away with a cold knife. All remaining traces of balsam are then to be removed from around the cover by means of a rag *just moistened* with methylated spirit, or, what is better, with a mixture of equal parts of spirit and æther, after which the slide is to be thoroughly washed in cold water. The slide is now in reality finished, but, in order to give it a smart appearance, it is usual, with the assistance of the turntable, to

run a ring of coloured varnish round the covering glass. A very useful varnish for the purpose is the *white zinc cement.* To prepare this, dissolve an ounce of gum dammar in an ounce of turpentine by the aid of heat. Take one dram of oxide of zinc and an equal quantity of turpentine; rub them up together in a mortar, adding the turpentine drop by drop, so as to form a creamy mixture perfectly free from lumps or grit. One fluid ounce of the dammar solution previously made must now gradually be added, the mixture being kept constantly stirred (*Frey*). The cement, when made, should be strained through a piece of fine muslin, previously wetted with turpentine, into a small wide-necked bottle, which, instead of having a cork or stopper, should be covered with a loose metal cap. Instead of a bottle, the varnish may be kept in one of the collapsible tubes used by artists; but though this plan is highly recommended by many, it is not without its disadvantages. If the varnish becomes thick by keeping, a few drops of turpentine or benzole well stirred in will soon reduce it to a suitable consistence for use.

PART II.

25. *Special Methods.*—Having in the preceding pages entered at some length into the general subject of section-cutting, it remains for us now to consider those special methods of preparation which the peculiarities of certain objects demand. In order to keep the bulk (and consequent price) of this manualette within due bounds, we shall, without further preface, proceed to the description of these methods, in doing which every endeavour will be made to employ such brevity of expression as may be consistent with perfect clearness of meaning. As the most convenient plan, the objects here treated of will be arranged in alphabetical succession.

26. *Bone.*—Both transverse and longitudinal sections should be prepared, the former being the prettier and most interesting. After prolonged maceration in water, all fat, etc., must be removed and the bone dried, when as thin a slice as possible is to be cut off in the desired

E

direction, by means of a very fine saw. If the
section so obtained be placed upon a piece of
smooth cork it may, with the aid of a fine file
and the exercise of care, be further reduced in
thickness. It is then to be laid upon a hone
moistened with water, and being pressed gently
and *evenly* down upon it with the tip of the
finger (protected, if necessary, by a bit of cork
or gutta-percha), it must be rubbed upon the
stone until the desired degree of thinness has
been attained. Finally, in order to remove
scratches and to polish the section, it should
be rubbed upon a dry hone of very fine
texture, or upon a strop charged with putty-
powder. After careful washing in several
waters the section must be allowed thoroughly
to dry, when it may be mounted by the *dry
method* in the following manner:—A ring of
gold-size must, by means of the turntable, be
drawn in the centre of a slide, and the slide put
away in a warm place for several days (the
longer the better), in order that the ring may
become perfectly dry and hard. When this has
been accomplished the section is to be placed
in the centre of the ring, and a covering circle
of the requisite size having been cleaned, this
must have a *thin* ring of gold-size applied round
its margin. The cover is now to be placed in
position and gently pressed down, a spring clip

being employed, if necessary, to prevent it from moving. In about twenty-four hours another layer of the varnish should be applied, and the slide afterwards finished in the manner already described (§ 24). The above method is also applicable to the preparation of sections of *teeth*, and also of *fruit-stones* and other hard bodies, which are incapable of being rendered soft enough for cutting.

As the process just described, however, is both troublesome and tedious, it is much better for ordinary purposes to have recourse to the *decalcifying method*, by which means sections in every way suitable for the examination of the essential structure of bone may be obtained with ease. To carry out this plan a piece of fresh bone should be cut into small pieces and placed in a solution made by dissolving 15 grains of pure chromic acid in 7 ounces of distilled water, to which 30 minims of nitric acid s.g. 1·420 are afterwards to be added. Here they should remain for three or four weeks, or until the bone has become sufficiently soft to cut easily, the fluid being repeatedly changed during the process. From this solution they must be transferred to methylated spirit for a few days, when a piece may be selected, imbedded in paraffine, and cut in the microtome (§ 12). Some of the sections should be

mounted, unstained, in spirit.   For this purpose
a cell of gold-size, as above described, must
first be prepared and filled *full* of a mixture of
spirit of wine one part, and distilled water
three parts.   Into this the section must be carefully placed and the cover applied, the same
precautions for the exclusion of air-bubbles
being taken which were recommended when
speaking of mounting in glycerine (§ 16).
When the cover is in position a ring of gold-
size must be laid on, repeated when dry, and
the slide afterwards finished in the ordinary
manner.   It will also be advisable to stain some
of the sections with carmine (§ 14), or picro-
carmine (§ 42), and mount them in glycerine.
Teeth may also be treated by the decalcifying
method, but in this case it must be remembered
that the enamel will dissolve away.

27. *Brain.*—The best hardening fluid is that
recommended by Rutherford, and is made by
dissolving 15 grains of pure chromic acid and
31 grains of crystallized bichromate of potash
in 43 ounces of distilled water.   Small pieces of
brain, which have previously been immersed
for twenty-four hours in rectified spirit, should
be placed in about a pint of this solution,
where they must remain for five or six weeks,
the fluid being repeatedly changed during the
process.   If by this time they are not suffi-

ciently hard the induration must be completed in alcohol. Sections are easily cut in the microtome by the paraffine method (§ 12). These may advantageously be stained in a solution of aniline blue, made by dissolving 1½ grain of aniline blue in 10 ounces of distilled water, and adding 1 drachm of rectified spirit (*Frey*). As this stain acts very rapidly two or three minutes' immersion will generally be found long enough. The sections must then be mounted in balsam (§ 23).

28. *Cartilage.*—The method to be employed in the preparation of cartilage will entirely depend upon the nature of the staining agent, to the action of which the sections are to be submitted. Thus, if the elegant *gold method* is to be followed, it is necessary that the cartilage should be perfectly fresh; whilst if any of the other staining agents are to be employed the tissue may have been previously preserved in alcohol. An excellent object on which to demonstrate the gold process is to be found in the articular cartilage of bone. It is a very easy matter to obtain from the butchers the foot of a sheep which has just been killed. The joint is to be opened, and the bones dissociated, when they will be seen to have their extremities coated with a white glistening membrane—this is the *articular cartilage.*

Exceedingly thin slices must be at once cut from it, and as only small sections are required, a sharp razor may be used for the purpose, the blade being either dry or simply wetted with distilled water. The sections as cut are to be transferred to a small quantity of a half per cent. solution of chloride of gold in a watch glass. Chloride of gold may be purchased in small glass tubes hermetically sealed, each tube containing 15 grains, and costing about 2s. If, however, the student requires only a small quantity of the staining fluid he need not be even at this small expense, for as photographers for the requirements of their art always keep on hand a standard solution of chloride of gold of the strength of one per cent., a little of this may readily be obtained, and diluted to the required degree. After the sections have been exposed to the action of the staining fluid for about ten minutes they may be transferred to a small beaker of distilled water, and exposed to diffused light for about twenty-four hours, when they must be mounted in glycerine (§ 16).

Sections of cartilage may also be examined, without being stained, in which case the field of the microscope should be only very feebly illuminated. Or carmine staining (§ 14) may be resorted to—these sections show well in glycerine, or if the staining be made very deep, even Canada balsam may be employed, and with fair results.

Microscopists are indebted to Dr. Frances Elizabeth Hoggan for the description of a new method of staining, which we have found especially suited to the treatment of cartilage. The agent employed is *iron*, and the process, which is very simple, is as follows. Two fluids are necessary—(1) tincture of steel; (2) a two per cent. solution of pyrogallic acid in alcohol. A little of the former is to be poured into a watch glass, and into this the sections, after having been previously steeped in alcohol for a few minutes, are to be placed. In about two minutes the iron solution is to be poured away and replaced by solution No. 2. In the course of a minute or two the desired depth of colour will have been produced, when the sections are to be removed, washed in distilled water, and mounted in glycerine. The results obtained by this process are very beautiful, the colour produced being a very fine neutral tint, of delightful softness. The process also answers admirably in the case of morbid tissues, and we have now in our possession some sections of ulcerated cartilage tinged by the iron method, in which the minute changes resulting from the ulcerative disintegration are brought out with wonderful distinctness.

As the structure of cartilage differs according

to its purpose and situation, the student will find his time profitably employed in a careful examination of the following forms (*a*) *hyaline* —articular and costal; (*β*) *yellow fibro-carti-lage*—epiglottis, or external ear; (*γ*) *cellular*— ear of mouse. Sections of the *intervertebral ligaments* should also be made, in which the different kinds of cartilage may be examined side by side with each other.

29. *Coffee Berry* affords sections of great beauty. The *unroasted* berry should be soaked for hours or days in cold water until sufficiently soft; then imbedded in paraffine, and cut in the microtome (§ 12), the section being made in the direction of the long axis of the berry. Put up in glycerine, or stain rather strongly with carmine, and mount in balsam. The same method of treatment may also be applied to other hard berries or *seeds*.

30. *Fat.*—Adipose tissue may be hardened in alcohol, cut in paraffine, and mounted in glycerine. If the tissue has been injected the sections may be mounted in balsam, and are then very beautiful objects, showing the ca-pillary network encircling the fat cells.

31. *Hair.*—Longitudinal sections are readily made by splitting the hair with a sharp razor. It is more difficult to cut the hair transversely. This, however, may easily be done in the fol- ·

lowing manner. The hairs having previously been well soaked in æther to remove all fatty matters, a sufficient number of them must be selected to form a bundle about the thickness of a crow quill. This bundle, after being tied at each extremity with a bit of thread, is to be immersed for several hours in strong gum (§ 18), to which a few drops of glycerine have been added. On removal, the bundle must be suspended by means of a thread attached to one end of it, in a warm place until sufficiently hard, when it is to be imbedded and cut in paraffine (§ 12). Each section, as cut, is to be floated off the knife into methylated spirit. From this it is with the aid of the spoon (§ 14) to be transferred to a slide, the spirit tilted off, a drop of absolute alcohol added, when, after a minute or two, this also is to be drained off, the section treated with clove oil, and the mounting completed as described in § 23.

32. *Horn* varies very much in consistence, in some instances having a cartilaginous character, whilst in others it is almost bony. In the latter case, sections will have to be ground down in the manner explained when speaking of bone (§ 26). Where the texture is less dense, recourse may be had to prolonged steeping in hot or boiling water; in some cases it will be necessary to continue the immersion

for several hours. When sufficiently soft the piece of horn may, by means of bits of soft wood, be firmly wedged into the tube of the microtome, and sections cut with a razor, or what is better, with a broad and very sharp chisel. The sections are to be put between glass slips, held together by American clips (or pegs), and put away for two or three days in order to become thoroughly dry. After well soaking in good turpentine or benzole, they must be transferred to slides, the superfluous turpentine drained off, and chloroform-balsam added, etc. (§ 23). Sections of horn should, of course, be cut in different directions, but for examination with the polariscope those cut transversely yield by far the most magnificent results. *Hoofs, whalebone,* and allied structures should also be treated by the above method.

33. *Intestine.*—The method to be pursued with *sections* has already been described (§ 18). The ileum, however, is a very pretty object when a portion of it is so mounted as to show the *villi erect.* To do this it is necessary to cement to the slide, by marine glue, a glass cell of sufficient depth. This should have been prepared some time beforehand, so that the cement may be perfectly dry and hard. The cell is now to be filled with turpentine, and the piece of ileum (having been previously passed

through methylated spirit and absolute alcohol into turpentine) is gently placed into it, having the villi uppermost; pour some pure and rather fluid balsam on the object at one end, and gradually incline the slide, so as to allow the turpentine to flow out at the opposite side of the cell, till it is full of balsam. Then take a clean cover, and having placed upon it a small streak of balsam from one end to the other, allow it gradually to fall upon the cell, so as to avoid the formation of air-bubbles (§ 17), and finish the slide in the usual manner.* Or, the intestine may be dried, and mounted *dry*, in a cell with a blackened bottom, for examination as an opaque object.

34. *Liver.*—Small pieces of liver may be very successfully hardened by immersion in alcohol, beginning with weak spirit and ending with absolute alcohol. Cut and mount as usual.

35. *Lung* must be prepared in chromic acid (§ 5). For the cutting of sections the freezing microtome (§ 18) is of especial value, and should, therefore, be used. If, however, the student be not provided with this instrument, he must proceed as follows. A small piece of lung, previously deprived of all spirit, is to be immersed until thoroughly saturated in solu-

* Ralf.

tion of gum (§ 18). A small mould of bibulous
paper (§ 2), only just large enough to receive
the piece of tissue, having been prepared and
filled with the mucilage, the specimen is to be
transferred to it. The mould, with its contents,.
is now to be placed in a saucer, into which a
mixture of about 6 parts of methylated spirit
and 1 part of water (*Schäfer*) is to be poured
until the fluid reaches to within about a third
of the top of the paper mould. In the course
of several hours the surface of the mucilage
will begin to whiten and solidify. As soon as
this occurs more dilute spirit must be poured
into the saucer, until the mould is completely
submerged. In a day or two the gum will be
found to have acquired a suitable consistence
for cutting, when it must be removed from the
spirit, the paper mould peeled off, and the mass
imbedded and cut in paraffine, the sections being
afterwards treated as if they had been obtained
by the freezing method (§ 18). If the solidifi-
cation of the gum should proceed too slowly, a
few drops of pure spirit may be added to the
contents of the saucer. If, on the other hand,
the gum should become overhard, it will be
necessary to put into the saucer a few drops
of water, and repeat this until the required
consistence be obtained.

36. *Muscle.*—Harden in chromic acid, and

cut in paraffine. Transverse sections may be
made to show the shape of the fibrils. Longitudinal sections will only be required in the
case of injected tissues, when such sections
will be found very elegant, showing, as they
do, the elongated meshes of capillaries running
between and around the muscular fasciculi.
Mount in glycerine or balsam. To see the
transverse striæ characteristic of voluntary
muscle, a very good plan is to take a bit of
pork (cooked or fresh), and by means of needles
to teaze it out into the finest possible shreds.
If these be examined in water or glycerine, the
markings will be shown very perfectly.

37. *Orange-peel*, common object though it
be, is not to be despised by the microscopist
Transverse sections must be prepared by the
gum method (§ 35). These sections are not to
be subjected to the action of alcohol (as this
would destroy the colour), but after *drying*
between glass slides they must be soaked in
turpentine and mounted in balsam. We shall
then have a good view of the large globular
glands whose office it is to secrete that essential oil upon which the odour of the orange
depends.

38. *Ovary* may be prepared in the same
manner as liver (§ 34). Sections, which are
to be cut in paraffine, may be stained with

carmine, and mounted in glycerine or balsam. Apart from all scientific value, we know of no slide for the microscope which, even as a mere object of show, surpasses in beauty a well-prepared section of *injected* ovary, showing the wondrous Graafian vesicles, surrounded by their meandering capillaries.

39. *Porcupine Quill.*—Soften in hot water, cut in paraffine, and mount in balsam. Much (in our opinion *too* much) lauded as an object for the polariscope.

40. *Potato.*—From the large amount of water which it contains thin sections cannot be cut from the potato in its natural state. It must, therefore, be partially desiccated, either by immersion in methylated spirit for a few days or by exposure to the air. Sections may then readily be obtained by imbedding and cutting in paraffine. Such sections mounted in balsam are very beautiful, the starch being seen *in situ*, whilst if polarized light be employed each granule gives out its characteristic black cross.

41. *Rush* is to be prepared and cut as orange-peel (§ 37). Transverse sections of this "weed" furnish slides of the most exquisite beauty.

42. *Skin.*—To prepare skin for section a piece is to be selected which, after having been boiled for a few seconds in vinegar, must be

stretched out on a bit of flat wood, and being maintained in position by pins be allowed to remain until thoroughly dry. Then imbed in paraffine, and cut *exceedingly* thin transverse sections. These may be stained in carmine, but more beautiful results are obtained if picro-carmine be employed. Sections of skin, when stained by this agent, are much increased both in beauty and instructiveness; for the several constituents of the tissue becoming tinged with different colours are readily distinguishable from each other, whilst the contrast of colouring forms a pleasing picture to the eye. The method of preparing picro-carmine is very simple, though it sometimes yields a solution not altogether satisfactory. The best formula with which we are acquainted is that given by Rutherford,* and if due care be taken in following it out failure will generally be avoided. " Take 100 c.c. of a saturated solution of picric acid. Prepare an ammoniacal solution of carmine, by dissolving 1 gramme in a few c.c. water, with the aid of excess of ammonia and heat. Boil the picric acid solution on a sand bath, and when boiling add the carmine solution. Evaporate the mixture to dryness. Dissolve the residue in 100 c.c. water, and filter. A clear solution

* " Practical Histology," 2nd edit. p. 173.

ought to be obtained; if not, add some more
ammonia, evaporate, and dissolve as before."
Sections may be exposed to the action of this
fluid for a period varying from fifteen to thirty
minutes, then rapidly washed in water, and
mounted in glycerine. They may also be
mounted in balsam, care being taken in that
case to shorten as much as possible the period
of their immersion in alcohol, so that no risk
may be run of the picric acid stain being
dissolved out.

If it is intended to study the structure of the
skin with anything like thoroughness, portions
must of course be examined from different
localities, in order that its several varieties and
peculiarities may be observed. Thus the *sudoriforous*, or sweat glands, may be found in the
sole of the foot, whilst the *sebaceous* glands are
to be sought in the skin of the nose. The
papillæ are well represented at the tips of
the fingers,* whilst the structure of the shaft
of the *hair*, together with that of the follicle
within which its root is enclosed, as also the

---

* It is well, in connection with these papillæ, to bear in
mind a fact pointed out by *Frey*, namely, that the tips of
the fingers frequently become, *post-mortem*, the seat of
extensive natural injections; hence, in sections from this
region, we frequently obtain good views of distended capillaries without having been at the trouble of previously
injecting them.—*Frey*, "Microscopical Technology."

muscles by which it is moved, are to be studied in sections of skin from the scalp or other suitable locality.

43. *Spinal Cord.*—The spinal cord, say of a cat or a dog (or if procurable, of man), after being cut into pieces about half an inch in length, may be hardened in the usual chromic acid fluid (§ 5). As it is peculiarly liable to overharden and become uselessly brittle, the process must be carefully watched. Its further treatment is the same as that of brain. These sections may be stained very satisfactorily by the *ink process*, for communicating details of which we are indebted to the kindness of Dr. Paul, of Liverpool. The agent usually employed is Stephenson's blue-black ink, which, for this purpose, must be quite fresh. As in the case of carmine, two methods of staining may be adopted—either rapid, by using concentrated solutions, or more prolonged, according to degree of dilution. For the reasons previously given (§ 14), slow methods of staining are always to be preferred, as yielding the most beautiful results, yet, for the purposes of preliminary investigation, it is often convenient to have recourse to the quick process. To carry out the latter plan, an ink solution of the strength 1 in 5—10 parts of water is to be freshly prepared, and the sections exposed

F

to its action for a few minutes. For gradual staining the dilution must be carried to 1 in 30—50, and the time of immersion prolonged to several hours, the sections being occasionally examined during the staining, so that they may be removed just as they have· acquired the desired tint. When a satisfactory coloration has been obtained, the preparations should be mounted in dammar or balsam (§ 23). One advantage of this method of staining is, that definition is almost as good by artificial light as by day.

44. *Sponge* may readily be cut after being tightly compressed between two bits of cork; or its interstices may be filled up by immersion either in melted paraffine (§ 11) or in strong gum (§ 18), and then cut as usual.

45. *Stomach* requires no special method of hardening (chromic acid). Sections should always, when practicable, be cut in the freezing microtome. In default of this, proceed in the manner as directed for lung (§ 35). Both vertical and horizontal sections will, of course, be required. If the preparation has been injected, the latter are particularly beautiful. Stain with carmine or aniline blue (§ 27), and mount—if for very close study, in glycerine— if injected and for a " show" slide, use balsam.

46. *Tongue.*—Harden in chromic acid, imbed and cut transverse sections in paraffine. As, however, the paraffine is apt to get entangled amongst the *papillæ*, whence it is afterwards with difficulty dislodged, it will be as well before imbedding to soak the tongue in strong gum for a *few minutes*, and afterwards immerse in alcohol till the gum becomes hardened, so that the delicate papillæ may thus be protected from the paraffine by a surface-coating of gum. The best staining agent is picro-carmine (§ 42). Sections of *cat's* tongue near the root, when thus stained, furnish splendid objects. Sections should also be made of the *taste-bulbs*, found on the tongues of rabbits. These are small oval prominences, situated one on each side of the upper surface of the tongue near its root. They should be snipped off with scissors, and vertical sections made in the direction of their long axis. Stain with carmine or picro-carmine, and mount in glycerine or balsam.

47. *Vegetable Ivory.*—After prolonged soaking in cold water may readily be cut in the microtome. The sections should be mounted in balsam, and though not usually regarded as polariscopic objects, nevertheless, when examined with the *selenite*, yield very good colours.

48. *Wood.*—Shavings of extreme thinness

may be cut from large pieces or blocks of timber, by means of a very sharp plane. In this way very good sections may be procured of most of the common woods, as oak, mahogany, "glandular wood" of pine, etc. Where, however, the material to be operated upon takes the form of stems, roots, etc., of no great thickness, they should, after having been reduced to a suitable consistence (§ 4), be imbedded in paraffine, and cut in the microtome. Before imbedding it must not be forgotten to immerse the wood to be cut in weak gum-water (§ 11), this precaution being of great importance, especially in the case of stems, etc., the bark of which is at all rough and sinuous. If the sections are to be mounted *unstained*, they are usually put up in weak spirit (§ 26). A very general method also of dealing with this class of objects is to mount them *dry* (§ 26). This plan, however, cannot be recommended, for however thin the sections may be, the outlines, when this process is adopted, always present a disagreeable black or blurred appearance. To avoid this we may have recourse to Canada balsam, but the ordinary method of employing it must be slightly modified, a drop of chloroform being substituted for the clove oil (§ 23), otherwise this latter agent will cause the section to become so transparent as to render minute details of structure difficult to recognize. A better plan, perhaps, is to stain the section with carmine or logwood, and mount in balsam by the ordinary process. The best course to follow, however, especially in the case of transverse sections, is the *double staining* method.* For this purpose the sections in the first place must be subjected to the action of a solution of chloride of lime ($\frac{1}{4}$ oz. to a pint of water) until they become thoroughly bleached. They must then be soaked in a solution of hyposulphite of soda (one drachm to four ounces of water) for an hour, and after being washed for some hours, in several changes of water, are to be transferred for a short time to methylated spirit. Some red staining-fluid is now to be prepared by dissolving half a grain of Magenta crystals in one ounce of methylated spirit. A little of this solution being poured into a small vessel of white porcelain (§ 14), the sections are to be immersed in the dye for about thirty minutes. They are now to be removed, and after *rapid* rinsing in methylated spirit to remove all superfluous colour, they must be placed in a blue staining fluid made by dissolving half a

* See a paper by Mr. Stiles in the "Pharmaceutical Journal;" also, "Monthly Microscopical Journal" for August, 1875.

grain of aniline blue in one drachm of distilled water, adding ten minims of dilute nitric acid, and afterwards sufficient methylated spirit to make two fluid ounces. The sections must be permitted to remain in this solution for a very short time only, one to three minutes being generally sufficient, for as the action of the dye is very energetic, it will, if too long exposure be allowed, completely obliterate the previous coloration by the magenta. After being again *rapidly* rinsed in methylated spirit, as much of this as possible must be drained off, and the sections put into oil of *cajeput*, whence, in an hour, they may be transferred to spirits of turpentine, and, after a short soaking, mounted in balsam.

If the student will carefully carry out the above process, his trouble will be amply repaid by the beautiful results obtained, for by its means he may, with ease, prepare for himself a series of slides of such value as to constitute a worthy addition to his cabinet.

The preceding list by no means represents all the objects, sections of which will be found interesting to the microscopic student. Such was not its purpose — had it been so, the enumeration might have been prolonged al-most indefinitely. The end in view was to bring under the notice of the reader only those substances the cutting of which is accompanied by difficulty; and even of this class the space at our disposal has been so limited that we have been unwillingly compelled to pass over many, and dwell only on such as possess a typical character.

INDEX.

Acetic acid, 39
Adipose tissue, 72
Æther microtome, 25
Air-bubbles, 43, 59, 63
Alcohol, 16, 18
   absolute, 56

Balsam, Canada, 40, 58
   to clean from slide, 63
Beale's carmine fluid, 36
Bell's cement, 46
Blue staining fluid, 69
Bone, 15, 65
   decalcified, 67
Bottles for media, &c., 58
Brain, 68
Brunswick black, 46

Capillary attraction, 63
Carmine staining, 35
Carpenter, Dr., 41, 46
Cartilage, 69
Cells, to make, 66
Centring slide, 42
Chloroform-balsam, 59, 61
Chromic acid, 16
Clove oil, 57, 62
Coffee berry, 72
Cover, applying, 44
      German plan, 46
Crochet-needle holders, 43

Double staining, 85
Dry mounting, 66, 75, 84

Fat, 72
Finishing slide, 63

Fibro-cartilage, 72
Freezing method, 47-53
Fresh tissues, to cut, 12-14
Fruit stones, 67

Gardner, Mr., 26, 34
Gelatine cement, 45
Glycerine, 40, 41
Gold staining, 69
Gum, imbedding in, 20, 76
   strong, 49
   weak, 29, 84

Hair, 72, 80
Hæmatoxylon, see Logwood
Hand-sections, 19
Hardening agents, 16
Hoggan, F. Elizabeth, Dr.,
   71
Hone, 27, 66
Horn, hoofs, &c., 15, 73
Hyaline cartilage, 72

Ice, for freezing, 50
Injections, to harden, 19
Ink staining, 81
Intestine, 47, 48, 74
Iron staining, 71

Kay, J. A., Mr., 59
Kidney, 17
Klein's lifter, 38
Knife, section, 25, 32

Leaves, &c., to cut, 11
Ligaments, intervertebral, 72
Lime, carbonate of, 41

Liver, 75
Logwood staining, 53
Lung, 47, 75

Martindale's logwood fluid, 54
Media, mounting, 39
Methylated spirit, *see* Alcohol
Microtome, 21
    choice of, 22
Microtome, employment of, 31
    freezing, 47, 48
    imperfection in, 23, 50
    Rutherford's, 34
Muscle, 76

Needles, microscopical, 43

Orange-peel, 77
Ovary, 77

Papillæ of finger, 80
Paraffine, adhering to sections, 29
    imbedding in, 20, 27
    mixture, 28
    rising in tube, 23
    shrinks in cooling, 28, 31
Paul, Dr., 81
Pathological specimens, 19, 71
Picro-carmine, 68, 79, 83
Porcupine quill, 78
Potato, 78
Preparation of animal tissues, 15
    vegetable tissues, 14
Prussian-blue pigment, 42

Rangoon oil, 13, 53
Razors, 20, 27, 31, 32
Rush, 78

Rutherford's microtome, 24, 34
    " Practical Histology," 48
Salicylic acid, 49
Salt, for freezing, 50
Schäfer's " lifter," 38
    " Practical Histology," 54
Scissors, section, 12
Sebaceous glands, 80
Section-cutter, *see* Microtome
Section-knife, 25
Sections, thickness of, 34
    to dehydrate, 57
    to transfer, 37
Seeds, 72
Skin, 78
Sperm oil, 49
Spinal cord, 81
Spirit lamp for balsam, 60
    weak, 40, 68
Sponge, 82
Spoon, section, 37, 61
Staining agents, 34
Strop, razor, 27, 31
Stomach, 82
Sudoriforous glands, 80

Taste-bulbs, 83
Teeth, 15, 67
Tongue, 83
Tubes, collapsible, 64

Valentin's knife, 12
Vegetable ivory, 83
    tissues, to prepare, 14
Vessels, porcelain, 36

Water-bath, cheap, 28
Whalebone, 74
White zinc cement, 64
Wood, 83

PARDON AND SON, PRINTERS, PATERNOSTER ROW.

SELECTION FROM THE LIST OF WORKS

PUBLISHED BY

J. & A. CHURCHILL.

——◦—

CHEMISTRY, INORGANIC and ORGANIC. With Experiments. By CHARLES L. BLOXAM, Professor of Chemistry in King's College, London; Professor of Chemistry in the Department for Artillery Studies, Woolwich. Third Edition. With 295 Engravings. 8vo, 16s.

By the same Author.

LABORATORY TEACHING: or, Progressive Exercises in Practical Chemistry, with Analytical Tables. Third Edition. With 89 Engravings. Crown 8vo, 5s. 6d.

FOWNES' MANUAL OF CHEMISTRY. Edited by HENRY WATTS, B.A., F.R.S. Twelfth Edition.
Vol. I.—Physical and Inorganic Chemistry. With 154 Engravings and Coloured Plate of Spectra. Crown 8vo, 8s. 6d.
Vol. II. —Chemistry of Carbon Compounds, or Organic Chemistry. With Engravings. Crown 8vo, 10s.

QUALITATIVE ANALYSIS. By C. REMIGIUS FRESENIUS. Edited by ARTHUR VACHER. Ninth Edition, with Coloured Plate of Spectra and 47 Engravings. 8vo, 12s. 6d.

By the same Author.

QUANTITATIVE ANALYSIS. Edited by ARTHUR VACHER. Seventh Edition. Vol. I. With 106 Engravings. 8vo, 15s.

A PRIMER OF CHEMISTRY, Including Analysis. By ARTHUR VACHER. 18mo, 1s.

NEW BURLINGTON STREET.

PUBLISHED BY J. & A. CHURCHILL.

PRACTICAL CHEMISTRY AND QUALITATIVE INOR-
GANIC ANALYSIS. By FRANK CLOWES, D.Sc. Lond.,
F.C.S. Lond. and Berlin. Second Edition. With 47 Engravings.
Crown 8vo, 7s. 6d.

PRACTICAL CHEMISTRY, Including Analysis. By JOHN
E. BOWMAN and C. L. BLOXAM. Seventh Edition. With 98
Engravings. Fcap 8vo, 6s. 6d.

A HANDBOOK OF MODERN CHEMISTRY, Inorganic and
Organic. For the use of Students. By CHARLES MEYMOTT
TIDY, M.B., F.C.S., Professor of Chemistry at the London
Hospital. 8vo, 16s.

HOW TO TEACH CHEMISTRY: Hints to Science Teachers
and Students. Six Lectures by EDWARD FRANKLAND, D.C.L.,
F.R.S., Summarised and Edited by GEORGE CHALONER, F.C.S.
With 47 Engravings. Crown 8vo, 3s. 6d.

THE FIRST STEP IN CHEMISTRY: A New Method for
Teaching the Elements of the Science. By ROBERT GALLOWAY,
Professor of Applied Chemistry in the Royal College of
Science for Ireland. Fourth Edition, with Engravings. Fcap
8vo, 6s. 6d.

By the same Author.

A MANUAL OF QUALITATIVE ANALYSIS. Fifth Edition,
with Engravings. Post 8vo, 8s. 6d.

HANDBOOK OF VOLUMETRIC ANALYSIS: or, the Quantitative Estimation of Chemical Substances by Measure
applied to Liquids, Solids, and Gases. By FRANCIS SUTTON,
F.C.S., Public Analyst for the County of Norfolk. Third
Edition. With 74 Engravings. 8vo, 15s.

NOTES FOR STUDENTS IN CHEMISTRY: Being a
Syllabus of Chemistry, compiled mainly from the Manuals of
MILLER, FOWNES-WATTS, WURZ, and SCHORLEMMER. By
ALBERT J. BERNAYS, Ph.D., Professor of Chemistry at St.
Thomas's Hospital. Sixth Edition. Fcap 8vo, 3s. 6d.

HANDBOOK OF CHEMICAL TECHNOLOGY. By RUDOLF
WAGNER, Ph.D., Professor of Chemical Technology at the
University of Wurtzburg. Translated and Edited from the
Eighth German Edition, with Extensive Additions, by WILLIAM
CROOKES, F.R.S. With 336 Engravings. 8vo, 25s.

NEW BURLINGTON STREET.

INTRODUCTION TO INORGANIC CHEMISTRY. By WM.
G. VALENTIN, F.C.S., Principal Demonstrator of Practical
Chemistry in the Royal School of Mines and Science Training
Schools, South Kensington. Third Edition. With 82 En-
gravings. 8vo, 6s. 6d.

By the same Author.

QUALITATIVE CHEMICAL ANALYSIS. Fourth Edition.
With 19 Engravings. 8vo, 7s. 6d.

Also

TABLES FOR THE QUALITATIVE ANALYSIS OF SIMPLE
AND COMPOUND SUBSTANCES, both in the Dry and Wet
Way. On indestructible paper. 8vo, 2s. 6d.

Also

CHEMICAL TABLES FOR THE LECTURE ROOM AND
LABORATORY. In five large sheets. 5s. 6d.

COOLEY'S CYCLOPÆDIA OF PRACTICAL RECEIPTS,
AND COLLATERAL INFORMATION IN THE ARTS,
MANUFACTURES, PROFESSIONS, AND TRADES : In-
cluding Pharmacy and Domestic Economy and Hygiéne. Sixth
Edition, Revised and Rewritten by Professor RICHARD V.
TUSON, F.I.C., F.C.S., assisted by several Scientific Contributors.
In Monthly Parts, each 2s. 6d.

ROYLE'S MANUAL OF MATERIA MEDICA AND THE-
RAPEUTICS. Sixth Edition. By JOHN HARLEY, M.D. With
139 Engravings. Crown 8vo, 15s.

THE ELEMENTS OF THERAPEUTICS. A Clinical Guide
to the Action of Drugs. By C. BINZ, M.D., Professor of Phar-
macology in the University of Bonn. Translated and Edited
with Additions, in Conformity with the British and American
Pharmacopœias, by EDWARD I. SPARKS, M.A., M.B. Oxon.,
formerly Radcliffe Travelling Fellow. Crown 8vo, 8s. 6d.

THE STUDENT'S GUIDE TO MATERIA MEDICA. Including the New Additions to the British Pharmacopœia. By
JOHN C. THOROWGOOD, M.D. Lond., Lecturer on Materia
Medica at the Middlesex Hospital. With Engravings. Fcap
8vo, 6s. 6d.

MATERIA MEDICA AND THERAPEUTICS: VEGETABLE KINGDOM. By CHARLES D. F. PHILLIPS, M.D. 8vo, 15s.

LECTURES ON PRACTICAL PHARMACY. By BARNARD S. PROCTOR, Lecturer on Pharmacy at the College of Medicine, Newcastle-on-Tyne. With 43 Engravings and 32 Plates, containing Fac-simile Prescriptions. 8vo, 12s.

PHARMACEUTICAL GUIDE TO THE FIRST AND SECOND EXAMINATIONS. By JOHN BARKER SMITH. Second Edition. Crown 8vo, 6s. 6d.

COMPANION TO THE BRITISH PHARMACOPŒIA. By PETER SQUIRE, late President of the Pharmaceutical Society, assisted by his Sons P. W. SQUIRE and A. H. SQUIRE. Eleventh Edition. 8vo, 10s. 6d.

A TOXICOLOGICAL CHART, Exhibiting at one view the Symptoms, Treatment, and Mode of Detecting the Various Poisons, Mineral, Vegetable, and Animal. To which are added concise Directions for the Treatment of Suspended Animation. By WILLIAM STOWE, M.R.C.S.E. Thirteenth Edition. Sheet, 2s.; Roller, 5s.

THE POCKET FORMULARY AND SYNOPSIS OF THE BRITISH AND FOREIGN PHARMACOPŒIAS : Comprising Standard and approved Formulæ for the Preparations and Compounds employed in Medical Practice. By HENRY BEASLEY. Tenth Edition. 18mo, 6s. 6d.

By the same Author.

THE DRUGGIST'S GENERAL RECEIPT-BOOK : Comprising a Copious Veterinary Formulary, Numerous Recipes in Patent and Proprietary Medicines, Druggists' Nostrums, &c.; Perfumery and Cosmetics; Beverages, Dietetic Articles and Condiments; Trade Chemicals; Scientific Processes; and an Appendix of Useful Tables. Eighth Edition. 18mo, 6s. 6d.

Also

THE BOOK OF PRESCRIPTIONS: Containing 3,107 Prescriptions collected from the Practice of the most eminent Physicians and Surgeons, English and Foreign. With an Index of Diseases and Remedies. Fifth Edition. 18mo, 6s. 6d.

**POISONS IN RELATION TO MEDICAL JURISPRUDENCE
AND MEDICINE.** By Alfred S. Taylor, M.D., F.R.S.,
Professor of Medical Jurisprudence to Guy's Hospital. Third
Edition, with 104 Engravings. Crown 8vo, 16s.

PRACTICAL PHARMACEUTICAL CHEMISTRY: An Explanation of Chemical and Pharmaceutical Processes; with the
Methods of Testing the Purity of the Preparations, deduced
from Original Experiments. By Dr. G. C. Wittstein. Translated from the Second German Edition by Stephen Darby.
18mo, 6s.

A MANUAL OF BOTANY: Including the Structure, Func-
tions, Classifications, Properties, and Uses of Plants. By
Robert Bentley, F.L.S., Professor of Botany, King's
College, and to the Pharmaceutical Society. Third Edition,
with 1,138 Engravings. Crown 8vo, 14s.

MEDICINAL PLANTS: being Descriptions with Original
Figures of the Principal Plants employed in Medicine, and an
Account of their Properties and Uses. By Robert Bentley,
F.L.S., Professor of Botany in King's College, and to the
Pharmaceutical Society; and Henry Trimen, M.B., F.L.S.,
Lecturer on Botany in St. Mary's Hospital Medical School.
In Monthly Parts, each containing 8 Coloured Plates, 5s. each.

THE MICROSCOPE AND ITS REVELATIONS. By W. B.
Carpenter, M.D., F.R.S. Fifth Edition, with more than 500
Engravings. Crown 8vo, 16s.

A MANUAL OF MICROSCOPIC MOUNTING; with Notes
on the Collection and Examination of Objects. By John H.
Martin, Member of the Society of Public Analysts, Author of
"Microscopic Objects." Second Edition. With 150 Engravings.
8vo, 7s. 6d.

THE MICROSCOPIST: A Manual of Microscopy and Com-
pendium of the Microscopic Sciences, Micro-Mineralogy, Micro-
Chemistry, Biology, Histology, and Pathological Histology.
By J. H. Wythe, A.M., M.D., Professor of Microscopy and
Biology in the San Francisco Medical College. Third Edition,
with 205 Engravings. 8vo, 18s.

A MANUAL OF ANIMAL PHYSIOLOGY. With Appendix
of Examination Questions. By John Shea, M.D., B.A. Lond.
With numerous Engravings. Fcap 8vo, 5s. 6d.

PUBLISHED BY J. & A. CHURCHILL.

A MANUAL OF THE ANATOMY OF INVERTEBRATED
ANIMALS. By Prof. HUXLEY, LL.D., F.R.S. With 158
Engravings. Fcap 8vo, 16s.

By the same Author.

A MANUAL OF THE ANATOMY OF VERTEBRATED
ANIMALS. With 110 Engravings. Fcap 8vo, 12s.

CHAUVEAU'S COMPARATIVE ANATOMY OF THE
DOMESTICATED ANIMALS. Translated from the Second
French Edition, and Edited by GEORGE FLEMING, F.R.G.S.,
Veterinary Surgeon, Royal Engineers. With 450 Engravings.
8vo, 31s. 6d.

NOTES ON COMPARATIVE ANATOMY: a Syllabus of a
Course of Lectures delivered at St. Thomas's Hospital. By
WILLIAM MILLER ORD, M.B. Lond., M.R.C.P., Assistant-
Physician to the Hospital, and Lecturer in its Medical School.
Crown 8vo, 5s.

THE STUDENT'S GUIDE TO ZOOLOGY: A Manual of
the Principles of Zoological Science. By ANDREW WILSON,
Author of "Elements of Zoology," and Lecturer on Zoology,
Edinburgh. With Engravings. Fcap 8vo, 6s. 6d.

NOTES ON NATURAL PHILOSOPHY: Lectures delivered
at Guy's Hospital, by G. F. RODWELL, F.R.A.S., Science Master
in Marlborough College. With 48 Engravings. Fcap 8vo, 5s.

A MANUAL OF PHOTOGRAPHIC MANIPULATION. By
LAKE PRICE. Second Edition, with numerous Engravings.
Crown 8vo, 6s. 6d.

A MANUAL OF PHOTOGRAPHY. By GEORGE DAWSON,
M.A., Ph.D., Lecturer on Photography in King's College,
London. Eighth Edition, with Engravings. Fcap 8vo, 5s. 6d.

MEDICAL LEXICON: A DICTIONARY OF MEDICAL
SCIENCE. By ROBLEY DUNGLISON, M.D. New Edition, by
RICHARD J. DUNGLISON, M.D. Royal 8vo (1,130 pp.), 28s.

MAYNE'S MEDICAL VOCABULARY: being an Explanation of all Names and Phrases used in the various depart-
ments of Medical Science and Practice, giving their Derivation,
Meaning, Application, and Pronunciation. Fourth Edition.
Fcap 8vo, 10s

NEW BURLINGTON STREET.